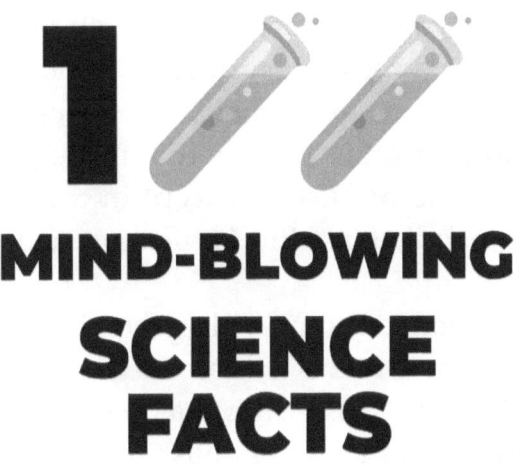

MIND-BLOWING
SCIENCE
FACTS

100 Incredible Discoveries That Changed
Our World

FELIX GRAYSON

MINDSPARK
PUBLISHING

CONTENTS

BEFORE WE DIVE IN...

Did you know that this is just **one** of many **mind-blowing** books waiting to be discovered?

What if I told you there's a **world of jaw-dropping, unbelievable, and downright bizarre facts** across **sports, science, history, mysteries, and more**—each one packed with stories that will **challenge what you thought you knew?**

EVER WONDERED WHAT IT'S LIKE TO...

- Witness **record-breaking Olympic moments** that defy human limits?

- Explore **real-life conspiracy theories** that sound too wild to be true?

- Discover **unsolved mysteries** that still leave experts baffled?

- Learn about **billionaires, stock market**

crashes, and money secrets?

- Find out how **robots, AI, and space travel are shaping the future?**

- Experience the **most extreme sports, legendary battles, and shocking events?**

This is just the beginning. The **100 Mind-Blowing series** covers it **all.**

WANT TO SEE WHAT'S NEXT?

Go to **FelixGrayson.com** and explore the **growing collection** of books and audiobooks that will **entertain, amaze, and keep you coming back for more.**

Curiosity doesn't stop here—this is just the beginning. What will blow your mind next?

INTRODUCTION

Welcome to *100 Mind-Blowing Science Facts*, a collection that's sure to make you say, "Wait, that's actually true?" From mind-boggling discoveries to unbelievable phenomena, this book is packed with jaw-dropping stories that will make you rethink everything you thought you knew about science.

Ever wondered what would happen if you could survive being frozen alive? Or how some fish can generate their own electricity? How about the planet where it rains glass or the jellyfish that can live forever? These are just a few of the unbelievable stories waiting for you inside. Each fact has been carefully chosen to surprise, entertain, and maybe even challenge what you believe is possible.

Whether you're here for a fun read, a conversation starter, or just looking for something to spark your curiosity, this book has something for everyone. Read it cover to cover, or flip to a random page and let your imagination run wild. There's no right or wrong way to enjoy this journey through the strange, fascinating, and mind-bending world of science.

So grab your favorite snack, get comfy, and prepare to explore some of the most mind-blowing moments in the history of science. Who knows? By the end, you might just have a few wild facts of your own to share. Let's dive in!

Mind-Blowing Science Fact #1

THE ACCIDENTAL DISCOVERY OF PENICILLIN

In 1928, Alexander Fleming changed the world—by forgetting to clean his lab!

When Fleming returned from vacation, he noticed that a mold had accidentally grown on one of his petri dishes. Strangely, the bacteria around the mold had been killed off. That mold turned out to be *Penicillium notatum*, and it led to the creation of **penicillin**—the world's first true antibiotic. This chance discovery has since saved over **200 million lives** and revolutionized modern medicine. All because of a messy workspace!

Mind-Blowing Science Fact #2

THE PLANET THAT RAINS DIAMONDS

On Neptune and Uranus, it literally rains diamonds!

These icy giant planets have extreme pressure and temperatures deep within their atmospheres. Scientists believe that this intense pressure crushes carbon atoms together, forming **solid diamonds** that fall like rain through the atmosphere. It's basically a cosmic jewelry storm happening 30 times farther from the sun than Earth. Talk about *precious weather*!

Mind-Blowing Science Fact #3

THE TREE THAT OWNS ITSELF

There's a tree in Georgia that legally owns itself!

In Athens, Georgia, an old white oak tree has one of the strangest legal statuses in the world. In the early 1800s, a man named William Jackson supposedly deeded the tree and the land around it **to itself**—meaning the tree is technically the owner of its own property. While the legality of the deed is questionable, the town has honored the claim for over a century. The current tree is actually a descendant of the original and is still known as **The Tree That Owns Itself**.

Mind-Blowing Science Fact #4

THE ANIMAL THAT CAN REGROW ITS BRAIN

Flatworms can literally regrow their own brains!

If you cut a flatworm in half, it doesn't just survive—it regenerates into **two fully functional flatworms**, complete with brand-new brains. Even crazier, scientists discovered that after regrowing their brains, these worms **retain memories** they learned before being sliced. It's one of the most mind-blowing examples of regeneration in the animal kingdom and has researchers dreaming about the future of human tissue repair.

Mind-Blowing Science Fact #5

THE COLDEST SPOT IN THE UNIVERSE

The coldest known place in the universe… is on Earth!

In 1995, scientists at the University of Colorado created something colder than deep space itself. Using lasers and magnetic fields, they cooled a tiny cluster of atoms to just **0.0000000001 Kelvin** — a fraction above absolute zero. This state is called a **Bose-Einstein Condensate**, where atoms behave like a single "super-atom" with bizarre quantum properties. Outer space might be cold, but human science has managed to go even colder.

Mind-Blowing Science Fact #6

THE BATTERY MADE OF POTATOES

You can power a lightbulb with a potato!

A simple potato can be turned into a working battery by inserting a **zinc nail** and a **copper coin** into it. The potato's natural acids act as an electrolyte, allowing electrons to flow and generate a small electric current. It's not exactly enough to power your phone, but it's enough to light up an LED. This classic science experiment proves that even the humblest vegetable can teach us the basics of electricity.

Mind-Blowing Science Fact #7

THE MAN WHO WEIGHED THE EARTH

A scientist once calculated Earth's weight... with a mountain!

In 1774, British scientist **Henry Cavendish** performed an experiment so precise, it's still impressive today. By using a torsion balance and measuring the tiny gravitational pull between lead spheres, he was able to calculate the mass of the entire **Earth**—without leaving his lab. His method became known as "**weighing the Earth**," and his result was remarkably close to modern measurements. All of this, without computers, satellites, or fancy tech.

Mind-Blowing Science Fact #8

THE JELLYFISH THAT LIVES FOREVER

There's a jellyfish that can literally live forever.

The **Turritopsis dohrnii**, also known as the "immortal jellyfish," has a unique ability: when it's injured or old, it can revert its cells back to their **youthful state** and start its life cycle all over again. This process, called **transdifferentiation**, allows it to cheat death indefinitely—making it the closest thing to biological immortality ever discovered.

Mind-Blowing Science Fact #9

THE GPS THAT RELIES ON EINSTEIN

Your GPS wouldn't work without Einstein's theories!

The Global Positioning System (GPS) relies on satellites orbiting Earth—and those satellites have to account for both **Special** and **General Relativity**. Time actually moves **faster** in orbit due to weaker gravity and **slower** because of their high speed. If engineers didn't adjust for these tiny time differences, your GPS would be off by **about 10 kilometers per day**! So every time you check directions, you're using Einstein's genius.

Mind-Blowing Science Fact #10

THE FROG THAT FREEZES SOLID

Some frogs can survive being frozen alive! The **wood frog** of North America has an incredible survival trick: when winter hits, it allows itself to **freeze solid**. Its heart stops, its blood turns to ice, and it appears completely dead. But when spring arrives, the frog **thaws out** and hops away as if nothing happened. It survives thanks to special proteins and glucose in its body that prevent its cells from bursting during the freeze.

Mind-Blowing Science Fact #11

THE ROBOT THAT PASSED THE TURING TEST

A computer once tricked humans into thinking it was real.

In 2014, a chatbot named **Eugene Goostman** became the first to officially pass the **Turing Test**—a challenge proposed by Alan Turing in 1950 to see if a machine could imitate human conversation well enough to fool people. Eugene pretended to be a 13-year-old Ukrainian boy and managed to convince **33% of the judges** that he was human. It wasn't perfect AI, but it was a historic milestone in machine intelligence.

Mind-Blowing Science Fact #12

THE SOUND OF BLACK HOLES

B lack holes can "sing" across the universe! In 2003, NASA discovered that a supermassive black hole in the **Perseus Galaxy Cluster** was producing sound waves. These waves translated into an incredibly deep **B-flat note**, about **57 octaves below middle C**—far too low for humans to hear. Technically, it's the **deepest sound ever detected** in the universe. So, somewhere out there, space itself is humming a cosmic tune.

Mind-Blowing Science Fact #13

THE MUSHROOM THAT BREATHES LIKE US

Some fungi breathe oxygen and exhale carbon dioxide!

While plants famously take in carbon dioxide and release oxygen, certain **mushrooms and fungi** do the exact opposite—they **breathe like animals**. They absorb oxygen and release carbon dioxide as they break down organic matter. In fact, the largest living organism on Earth is a **fungus** in Oregon, covering over **2,300 acres** underground and breathing quietly beneath the forest floor.

Mind-Blowing Science Fact #14

THE PLANET MADE OF PURE CRYSTAL

There's a planet that's basically a giant diamond!

Astronomers discovered a planet called **55 Cancri e** that's believed to be made mostly of **carbon**—and because of its extreme heat and pressure, that carbon is likely in the form of **crystalline diamond**. This alien world is twice the size of Earth and orbits so close to its star that a year there lasts only **18 hours**. Basically, it's a planet-sized gemstone orbiting in space.

Mind-Blowing Science Fact #15

THE BLOOD THAT'S BLUE AND GREEN

Not all animals have red blood like us! While humans and most animals have **red blood** because of iron-rich hemoglobin, some creatures use completely different chemistry. **Octopuses, squids, and horseshoe crabs** have **blue blood** thanks to a copper-based molecule called **hemocyanin**. Even stranger, some New Guinea lizards have **green blood** due to a buildup of bile pigment. Nature has a whole rainbow hidden inside.

Mind-Blowing Science Fact #16

THE MOON SMELLS LIKE GUNPOWDER

Astronauts say the Moon has a smell!

When Apollo astronauts returned to their lunar module after moonwalks, they noticed a distinct scent clinging to their suits and equipment. They described it as smelling like **burnt gunpowder**. Lunar dust, or **regolith**, has no smell in space (since space is a vacuum), but once exposed to the air inside the spacecraft, it reacted strangely—leaving that smoky, metallic scent hanging in the cabin.

Mind-Blowing Science Fact #17

THE BACTERIA THAT EAT METAL

Some bacteria literally feed on metal!

Scientists have discovered certain types of bacteria that can **consume and "digest" metals** like iron and manganese. These microbes use the metals as an **energy source**, stripping away electrons in a process called **metal oxidation**. Researchers are even exploring ways to use these bacteria to help clean up polluted environments or recycle toxic waste—proof that life can thrive in the most unexpected ways.

Mind-Blowing Science Fact #18

THE LIGHTNING THAT STRIKES TWICE

Lightning can—and often does—strike the same place twice!

Despite the old saying, lightning isn't shy about hitting the same spot multiple times. In fact, the **Empire State Building** in New York is struck by lightning about **20 to 25 times per year**! Tall structures, isolated trees, and even people standing in open fields are all prime targets for repeat strikes. Nature doesn't care about old sayings—it follows the rules of physics.

Mind-Blowing Science Fact #19

THE COLOR YOU CAN'T SEE

There's a color your brain can't process.

Called **forbidden colors**, these are pairs of hues like **red-green** or **blue-yellow** that the human eye physically can't see together because of how our color receptors work. Scientists have tried to create conditions where people might glimpse these impossible colors, but for most of us, they exist only in theory—a reminder that even something as simple as color has limits we can't naturally experience.

Mind-Blowing Science Fact #20

THE ANIMAL THAT SEES MAGNETISM

Some animals can literally see Earth's magnetic field!

Certain species, like **European robins** and **sea turtles**, have a biological ability called **magnetoreception**. Their eyes contain special proteins that allow them to actually **see magnetic fields** as visual patterns or shades of light, helping them navigate vast distances with pinpoint accuracy. It's like having a built-in compass—one that we humans can't even begin to imagine.

Mind-Blowing Science Fact #21

THE STAR THAT OUTSHINES GALAXIES

One star can shine brighter than entire galaxies!

Astronomers have discovered a type of star called a **hypergiant,** and one of the most extreme is **R136a1.** This single star is about **8.7 million times brighter than our Sun** and burns so hot and bright that it can outshine **small galaxies** from millions of light-years away. It's a cosmic monster that defies the limits of what we thought stars could do.

Mind-Blowing Science Fact #22

THE ANT THAT EXPLODES ITSELF

Some ants sacrifice themselves by exploding!

In Southeast Asia, there's a species called **Colobopsis saundersi**, also known as the **exploding ant**. When threatened, these ants **rupture their own bodies**, releasing a sticky, toxic goo that traps and kills predators. It's the ultimate act of self-sacrifice, all to protect their colony. Nature's tiniest warriors aren't afraid to go out with a bang—literally.

Mind-Blowing Science Fact #23

THE LIQUID THAT CLIMBS WALLS

Water can actually defy gravity!

Thanks to a phenomenon called **capillary action**, water and other liquids can **climb up narrow spaces** without any external force. It's why water rises inside a plant stem or why a paper towel soaks up a spill. Molecules stick to the sides of small tubes and pull each other upward—essentially letting the liquid **crawl against gravity**. Plants literally depend on this quiet superpower to survive.

Mind-Blowing Science Fact #24

THE GLASS THAT'S TECHNICALLY A LIQUID

The windows in old buildings are slowly melting!

If you visit ancient cathedrals, you might notice that their stained glass windows are often **thicker at the bottom**. That's because **glass isn't a true solid**—it's actually a **super-cooled liquid** that flows extremely slowly over centuries. The effect is almost imperceptible, but over hundreds of years, gravity subtly pulls the glass downward, changing its shape.

Mind-Blowing Science Fact #25

THE ANIMAL WITH THREE HEARTS

Octopuses have not one, but **three hearts**! Two of an octopus's hearts pump blood to its gills, while the third pumps blood to the rest of its body. Even weirder, when an octopus **swims**, the heart that serves its body **actually stops beating**—which is one reason these creatures prefer crawling to swimming. Oh, and by the way, their blood is **blue** because it's copper-based, not iron-based like ours.

Mind-Blowing Science Fact #26

THE FIRE THAT BURNS WITHOUT FLAMES

Fire doesn't always need flames to burn.

There's a type of combustion called **smoldering**, where materials like **peat, coal, or wood** burn **without visible flames**. Instead, they burn slowly, at lower temperatures, and can last for **months—or even years—underground**. Some of the longest-burning fires on Earth, like the **Centralia coal fire** in Pennsylvania, have been smoldering **since 1962** and are still going today.

Mind-Blowing Science Fact #27

THE BUTTERFLY THAT TASTES WITH ITS FEET

Butterflies can taste with their feet!

A butterfly's taste sensors aren't in its mouth—they're located on its **feet**. When a butterfly lands on a plant, tiny chemical receptors in its feet detect whether the plant is edible or suitable for laying eggs. It's like if you could step on your dinner and instantly know if it was delicious.

Mind-Blowing Science Fact #28

THE WATER THAT CAN BOIL AND FREEZE

Water can boil and freeze at the same time! This strange phenomenon is called the **triple point**—the exact temperature and pressure where **solid, liquid, and gas phases** of a substance coexist in perfect balance. For water, this happens at **0.01°C** and a specific low pressure. In a lab, you can literally watch water **freeze, boil, and remain liquid—all at once.** Nature loves to bend the rules when the conditions are just right.

Mind-Blowing Science Fact #29

THE BONE THAT'S STRONGER THAN STEEL

Your bones are tougher than you think!

Human bone is an engineering marvel—it's **five times stronger than steel** when compared by weight. The internal structure of bone, called **trabecular bone**, is a honeycomb-like lattice that makes it incredibly strong yet lightweight. Pound for pound, it can withstand more pressure than concrete or steel beams, all while constantly repairing itself throughout your life.

Mind-Blowing Science Fact #30

THE VIRUS THAT'S OLDER THAN HUMANS

Some viruses predate human existence!

The **Mimivirus**, one of the largest viruses ever discovered, is so complex that it blurs the line between living and non-living. Even more mind-blowing, scientists have found **giant viruses** frozen in **30,000-year-old Siberian permafrost**—and when thawed, they were still **infectious**. These ancient microscopic time capsules existed long before modern humans walked the Earth.

Mind-Blowing Science Fact #31

THE ANIMAL THAT CAN SURVIVE SPACE

Tiny creatures called tardigrades can survive in space!

Also known as **water bears**, tardigrades are nearly indestructible. They can survive **extreme heat, freezing cold, radiation, crushing pressure, and even the vacuum of space**. In 2007, scientists sent tardigrades into orbit, exposed them to open space—and they came back **alive**. These microscopic survival machines might outlast every other lifeform on Earth.

Mind-Blowing Science Fact #32

THE METAL THAT MELTS IN YOUR HAND

There's a metal that melts at body temperature!

Gallium is a silvery metal that looks like something from a sci-fi movie, but if you hold it in your hand, it will slowly **melt into a liquid**. Its melting point is just **29.8°C (85.6°F)**—slightly above room temperature. Scientists love using it for experiments, and it's even used in electronics and solar panels. Just don't try to make spoons out of it—they'll literally melt at the dinner table.

Mind-Blowing Science Fact #33

THE PLANT THAT COUNTS THE DAYS

Some plants can measure time without a clock!

Photoperiodism is a natural ability that lets plants sense the **length of day and night**. This is how flowers like **chrysanthemums** and crops like **rice** know exactly **when to bloom** or **when to grow**, based solely on seasonal light changes. They don't have brains, but they have built-in biological calendars more precise than most humans'!

Mind-Blowing Science Fact #34

THE LIQUID METAL THAT KILLED A PLANET

Mercury is the only metal that's liquid at room temperature!

Unlike most metals, **mercury** stays in liquid form even when it's cool, flowing like molten silver. But this beautiful element is also **extremely toxic**—so toxic, in fact, that it's believed to have contributed to the fall of ancient Rome. Roman aristocrats used mercury in makeup, medicines, and even food containers, unknowingly poisoning themselves over generations.

Mind-Blowing Science Fact #35

THE FISH THAT CREATES ELECTRICITY

Some fish can generate their own electricity! The **electric eel** can produce powerful electric shocks of up to **600 volts** to stun prey or defend itself. Its body is packed with specialized cells called **electrocytes** that act like tiny batteries. Even more impressive, electric eels can also emit weaker electric pulses to **navigate, communicate, and sense their surroundings**—basically turning themselves into living power plants.

Mind-Blowing Science Fact #36

THE VOLCANO THAT CREATES NEW LAND

Some volcanoes literally build new islands! In places like **Iceland** and **Hawaii**, volcanic eruptions don't just destroy—they **create**. When underwater volcanoes erupt, the cooled lava piles up until it breaks the surface, forming entirely **new islands**. One famous example is **Surtsey**, an island off Iceland that emerged in **1963** and became a living laboratory for scientists to watch how life colonizes brand-new land.

Mind-Blowing Science Fact #37

THE BRAIN CELLS THAT FIRE TOGETHER

Your brain rewires itself every day!

The human brain has a remarkable ability called **neuroplasticity**—the capacity to **reorganize and form new neural connections** throughout your life. Every time you learn something new or practice a skill, your brain's wiring literally changes. The classic phrase **"neurons that fire together, wire together"** is true: your thoughts and habits physically reshape your brain.

Mind-Blowing Science Fact #38

THE SHADOW THAT'S ALWAYS FASTER

Your shadow can move faster than light!

If you project a shadow onto a distant surface, the tip of that shadow can **travel faster than the speed of light** as you move the object. But here's the catch: no actual **information or matter** is traveling that fast—just the **absence of light**. So, while your shadow might break cosmic speed limits, it doesn't violate the laws of physics.

Mind-Blowing Science Fact #39

THE ICE THAT'S HOTTER THAN FIRE

There's an ice that stays solid at extreme heat!

Called **Ice VII**, this exotic form of ice doesn't melt at room temperature—it forms only under **enormous pressure**, like deep inside planets. Even when heated to **over 1,000°F (537°C)**, it can remain completely **solid** if the pressure is high enough. Scientists believe Ice VII exists in the deep interiors of **giant exoplanets** across the galaxy.

Mind-Blowing Science Fact #40

THE PLANT THAT TRAPS TIME

Some seeds can survive for thousands of years!

In 2007, scientists successfully **revived a 32,000-year-old seed** from an ancient squirrel burrow in Siberia. The plant, called **Silene stenophylla**, grew, flowered, and produced new seeds—all from a seed frozen since the Ice Age. Seeds like this can remain dormant for millennia, waiting patiently for the right conditions to spring back to life.

Mind-Blowing Science Fact #41

THE STAR THAT EATS PLANETS

Some stars can swallow their own planets!

Astronomers have observed dying stars known as **red giants** that expand so massively, they **engulf and consume** nearby planets. When a star like our Sun runs out of fuel, it swells up to hundreds of times its original size—sometimes **devouring entire planetary systems** in the process. It's a fiery reminder that nothing in space is permanent, not even planets.

Mind-Blowing Science Fact #42

THE RIVER THAT BOILS ITSELF

There's a real river that boils naturally!

Deep in the Amazon rainforest of Peru flows the **Shanay-Timpishka**, also known as the **Boiling River**. Its waters reach temperatures up to **200°F (93°C)** — **hot enough to cook animals** that fall in. Surprisingly, this isn't near a volcano; the heat comes from deep **geothermal activity** beneath the Earth's crust, making it one of the most mysterious and dangerous rivers in the world.

Mind-Blowing Science Fact #43

THE SHARK THAT GLOWS IN THE DARK

Some sharks naturally glow underwater!

Deep in the ocean, species like the **chain catshark** and **swell shark** produce a soft, greenish **bioluminescent glow**. But here's the twist—it's not just light; it's caused by **bioflu-orescence**, where the shark's skin absorbs blue light and re-emits it as green. Scientists believe they use this glow to **communicate** and **cam-ouflage** in the deep sea's darkness.

Mind-Blowing Science Fact #44

THE PARTICLE THAT CAN PASS THROUGH EARTH

Millions of particles fly through you every second!

They're called **neutrinos**—tiny, nearly massless particles that barely interact with matter. Every second, **about 100 trillion neutrinos** pass straight through your body without you ever noticing. They travel at near light-speed, coming from the **Sun, exploding stars, and even the Big Bang itself**. To them, Earth is almost completely transparent.

Mind-Blowing Science Fact #45

THE METAL THAT CAN FLOAT ON WATER

There's a metal so light, it floats!

Lithium and **sodium** are metals that can float on water because they're less dense than the liquid they're in. But here's the kicker: when these metals touch water, they don't just float—they **react violently**, often bursting into flames or even exploding due to a rapid chemical reaction with water. Lightweight, but seriously dangerous.

Mind-Blowing Science Fact #46

THE DESERT THAT FREEZES AT NIGHT

Some deserts get freezing cold after dark! Deserts like the **Sahara** aren't always scorching hot. Because deserts have **very little moisture** in the air and no cloud cover, they lose heat rapidly at night. Temperatures can drop from over **100°F (38°C)** during the day to **below freezing** after sunset. So, while we picture deserts as oven-like, they can also become icy, barren landscapes overnight.

Mind-Blowing Science Fact #47

THE ATOM THAT'S MOSTLY EMPTY SPACE

Most of everything is empty space! If you could zoom in on an atom, you'd find that it's almost entirely **nothing**. The nucleus is incredibly small compared to the space around it, and the electrons orbiting it are tiny. In fact, if an atom were the size of a stadium, the nucleus would be the size of a **pea** in the middle. That means everything around you—including you—is made of **99.9999999% empty space**.

Mind-Blowing Science Fact #48

THE RAIN THAT'S MADE OF ACID

Some rain is literally acidic!

Acid rain forms when pollutants like **sulfur dioxide** and **nitrogen oxides** mix with moisture in the atmosphere, creating **sulfuric and nitric acid**. When this acidic water falls to Earth, it can **damage forests, poison lakes, and erode buildings**. It's not acidic enough to burn skin, but over time, it quietly eats away at entire ecosystems and cities.

Mind-Blowing Science Fact #49

THE BEE THAT CAN DO MATH

Bees can understand zero and basic math! Studies have shown that **honeybees** can grasp the concept of **zero**—a tricky idea even for some humans—and can learn to **add and subtract** small numbers. Despite having a brain the size of a sesame seed, bees can process complex information, make decisions, and even recognize human faces. Tiny brains, but serious brainpower.

Mind-Blowing Science Fact #50

THE LAKE THAT TURNS TO BLOOD

There's a lake that naturally turns red!

Lake Natron in Tanzania is famous for its eerie, blood-red color. The lake's water is highly **alkaline** and rich in **salt-loving microorganisms** that produce red pigments. It's so salty and caustic that it can **mummify animals** that fall in. Despite its harshness, flamingos thrive there, using the toxic lake as a safe breeding ground away from predators.

Mind-Blowing Science Fact #51

THE RAIN THAT SMELLS LIKE EARTH

There's a scientific reason rain smells so good!

That fresh, earthy scent after a rainstorm is caused by a molecule called **geosmin**, released by soil-dwelling bacteria when rain hits the ground. Our noses are incredibly sensitive to geosmin—we can detect it at concentrations as low as **5 parts per trillion**. So that refreshing smell isn't just rain; it's **bacteria in the soil saying hello.**

Mind-Blowing Science Fact #52

THE RIVER THAT FLOWS BACKWARD

Once, a river in the U.S. flowed in reverse! In **1811 and 1812**, a series of powerful earthquakes along the **New Madrid Fault** caused the **Mississippi River** to temporarily flow **backward**. The quakes were so strong they created massive ground shifts, changed the landscape, and even formed **Reelfoot Lake** in Tennessee. It's one of the only times in recorded history a river has been seen reversing its course.

Mind-Blowing Science Fact #53

THE GLASS THAT CAN CUT STEEL

Some glass is harder than steel!

Volcanic glass, also known as **obsidian**, is formed when lava cools rapidly without crystallizing. Ancient cultures used it to craft razor-sharp tools and weapons. In fact, obsidian blades can be sharpened to an edge just a **few nanometers thick—sharper than surgical scalpels** and capable of slicing steel. Today, obsidian is even used in specialized surgical procedures.

Mind-Blowing Science Fact #54

THE OCEAN WITHOUT A SHORE

There's an ocean trapped inside the Earth! Scientists have discovered a massive reservoir of **water three times larger than all surface oceans combined**, hidden **400 miles beneath Earth's surface**. But it's not in liquid form—it's locked inside a blue mineral called **ringwoodite**. This underground ocean may play a key role in Earth's **tectonic activity** and **water cycle**, quietly shaping the planet from below.

Mind-Blowing Science Fact #55

THE FRUIT THAT CONDUCTS ELECTRICITY

Some fruits can generate electricity!

Citrus fruits like **lemons** and **oranges** can act as tiny **batteries** thanks to their acidic juice. When you insert a **zinc nail** and a **copper coin** into the fruit, the acid creates an **electrochemical reaction** that produces a small electric current. It's the same principle as the classic potato battery—but with a fruity twist.

Mind-Blowing Science Fact #56

THE ROCK THAT FLOATS ON WATER

Some rocks can actually float!

Pumice is a type of volcanic rock that's so full of **air pockets** and tiny holes that it can **float on water**. It forms during explosive volcanic eruptions when lava cools rapidly, trapping gas bubbles inside. Massive **pumice rafts** have even been spotted drifting across oceans, sometimes covering hundreds of square miles.

Mind-Blowing Science Fact #57

THE FLOWER THAT SMELLS LIKE ROT

Some flowers smell like rotting flesh!

The **corpse flower** (*Amorphophallus titanum*) produces one of the **worst smells on Earth**—a stench similar to decaying meat. It does this to attract **carrion beetles and flesh flies**, which help pollinate the plant. The flower can grow over **10 feet tall** and only blooms every few years, making its horrifying scent a rare but unforgettable event.

Mind-Blowing Science Fact #58

THE JELLY THAT NEVER MELTS

Some jellyfish can survive boiling heat!

The **Polyorchis penicillatus** jellyfish has evolved proteins that allow it to survive in **hydrothermal vents**, where water temperatures can exceed **750°F (400°C)**. These jellyfish live deep in the ocean, where they endure crushing pressure and scalding heat, making them some of the most **heat-tolerant animals** on the planet.

Mind-Blowing Science Fact #59

THE CLOUD THAT WEIGHS A MILLION POUNDS

Clouds are heavier than they look!

The average **cumulus cloud** (the fluffy, white kind) can weigh over **1 million pounds**. That weight comes from the **water droplets** suspended inside the cloud. So why don't clouds fall? Because the droplets are spread out over a massive area, and **air currents** keep them floating—making these giant, heavy water carriers drift effortlessly above our heads.

Mind-Blowing Science Fact #60

THE ANIMAL THAT CAN PAUSE TIME

Some animals can literally hit pause on their lives!

The **African lungfish** can survive **years without water** by entering a state called **estivation**. When their ponds dry up, they bury themselves in mud and slow their metabolism to near zero, effectively **pausing their biological clock**. They can stay like this for up to **4 years**, waiting for rain to return so they can wake up and swim again.

Mind-Blowing Science Fact #61

THE STORM THAT NEVER ENDS

There's a place where lightning never stops! Over **Lake Maracaibo** in Venezuela, a phenomenon called **Catatumbo lightning** occurs almost **300 nights a year**. Massive storm clouds unleash **up to 28 lightning strikes per minute**, making it the most **persistent lightning storm** on Earth. Scientists believe it's caused by the unique geography and warm air currents in the region.

Mind-Blowing Science Fact #62

THE FUNGUS THAT CONTROLS MINDS

There's a fungus that turns insects into zombies!

The **Ophiocordyceps** fungus infects ants and hijacks their brains, forcing them to climb to high places before killing them and sprouting out of their bodies. This creepy strategy allows the fungus to **spread its spores** far and wide. It's one of the most chilling examples of **mind control in nature**—real-life zombie behavior in the insect world.

Mind-Blowing Science Fact #63

THE PLANET THAT SPINS BACKWARD

One planet spins the opposite way!

Venus rotates in the **opposite direction** of most planets in our solar system. On Venus, the Sun would appear to **rise in the west and set in the east**. Even stranger, its day is longer than its year—it takes **243 Earth days to rotate once**, but only **225 Earth days to orbit the Sun**. Basically, a day on Venus lasts longer than a Venusian year.

Mind-Blowing Science Fact #64

THE BLOOD THAT CAN FREEZE WITHOUT DYING

Some fish have antifreeze in their blood!

In the icy waters of the **Antarctic**, certain fish species like the **Antarctic icefish** produce special **antifreeze proteins** in their blood. These proteins prevent deadly ice crystals from forming inside their bodies, allowing them to survive in temperatures that would freeze most other creatures solid. Nature's very own frost protection.

Mind-Blowing Science Fact #65

THE MOUNTAIN THAT GROWS EVERY YEAR

Mount Everest is still growing!

The mighty **Mount Everest** rises a little higher each year—by about **4 millimeters**—due to the ongoing **collision of tectonic plates** beneath it. The **Indian Plate** is slowly pushing up against the **Eurasian Plate**, forcing the Himalayas to continue rising. So technically, the world's tallest mountain keeps breaking its own record.

Mind-Blowing Science Fact #66

THE ISLAND MADE OF TRASH

There's a floating island of garbage in the ocean!

The **Great Pacific Garbage Patch** is a massive collection of **plastic waste and debris** trapped by ocean currents in the North Pacific. It covers an area estimated to be **twice the size of Texas** and contains millions of tons of plastic. It's one of the largest examples of how human activity can reshape entire ecosystems—even creating artificial "islands."

Mind-Blowing Science Fact #67

THE COLOR THAT DOESN'T EXIST

Your brain invents the color magenta! Unlike other colors, **magenta** isn't part of the visible light spectrum. There's no wavelength for it. Instead, your brain creates magenta by blending **red and blue light** and filling in the gap—since there's no green in between. Technically, magenta is a **psychological color**, a trick of perception your brain uses to make sense of light.

Mind-Blowing Science Fact #68

THE ANIMAL THAT BREATHES THROUGH ITS SKIN

Some animals can breathe without lungs! Creatures like **frogs, salamanders, and earthworms** can absorb oxygen directly through their **skin**. This process, called **cutaneous respiration**, allows them to breathe even underwater or underground. Some species, like the **barred tiger salamander**, rely almost entirely on skin breathing at certain stages of their lives.

Mind-Blowing
Science Fact #69

THE PLACE WHERE ROCKS MOVE THEMSELVES

Some rocks can slide across the desert!

In California's **Death Valley**, there's a mysterious phenomenon called **sailing stones**. Massive rocks, some weighing hundreds of pounds, leave long trails behind them as they **move across the dry lake bed**—seemingly on their own. Scientists discovered that thin layers of ice, combined with strong winds, gently push the rocks over time, leaving behind their famous tracks.

Mind-Blowing Science Fact #70

THE ANIMAL WITH SQUARE POOP

Wombats poop in perfect cubes!

Australian **wombats** are the only animals known to produce **cube-shaped poop**. They use these little blocks to **mark their territory**, and the shape prevents the poop from rolling away on uneven terrain. Scientists believe it's due to the unique elasticity of their intestines, which compress the waste into **squarish segments** before it exits.

Mind-Blowing Science Fact #71

THE PLANET THAT RAINS GLASS

On some planets, it rains sideways glass!

The exoplanet **HD 189733b** has some of the **most extreme weather** ever discovered. Winds there blow at over **5,400 mph (8,700 km/h)**, and it rains **molten glass—sideways—** because of the intense wind speeds. The planet's blue color, once thought to be like Earth's oceans, actually comes from light scattering off these violent, glassy storms.

Mind-Blowing Science Fact #72

THE FLOWER THAT TRICKS BEES

Some flowers pretend to be female bees!

Ophrys orchids, also known as **bee orchids**, have evolved to mimic the **appearance and scent** of female bees. Male bees mistake the flower for a mate and attempt to **mate with it**—a behavior called **pseudocopulation**. In the process, they unknowingly pick up and transfer pollen, helping the flower reproduce through pure deception.

Mind-Blowing Science Fact #73

THE LAKE THAT EXPLODES

Some lakes can suddenly explode!

In places like **Cameroon**, deep volcanic lakes such as **Lake Nyos** can build up massive amounts of **carbon dioxide gas** beneath the surface. When the pressure suddenly releases, the lake can **erupt in a deadly gas cloud**, suffocating everything nearby. In 1986, one such explosion at Lake Nyos tragically killed over **1,700 people and thousands of animals** in minutes.

Mind-Blowing Science Fact #74

THE SPIDER THAT BUILDS DECOYS

Some spiders craft fake spiders!

The **Cyclosa spider** not only builds webs but also creates **decoy spiders** out of leaves, debris, and dead insects. These fake spiders are often **larger than the real one** and serve as a distraction to predators. When something approaches, the real spider hides near its decoy, fooling attackers into striking the wrong target.

Mind-Blowing Science Fact #75

THE STAR THAT SPINS LIKE A BLENDER

Some stars spin at dizzying speeds!

Neutron stars, the collapsed cores of massive stars, can spin incredibly fast—up to 700 times per second. These spinning stars, called pulsars, emit beams of radiation that sweep across space like a cosmic lighthouse. If you could stand on one (you couldn't—it's impossibly dense), you'd be whipped around at a speed close to one-tenth the speed of light.

Mind-Blowing Science Fact #76

THE ICE THAT BURNS

There's ice that can catch fire!

Methane hydrate, also known as **flammable ice**, is a crystalline substance where **methane gas** is trapped inside a cage of **water molecules**. It looks like regular ice but can **ignite when exposed to flame**. Found deep under the ocean floor and in Arctic permafrost, it's considered one of the largest untapped energy sources on Earth—though risky to extract.

Mind-Blowing Science Fact #77

THE ANIMAL THAT CAN CLONE ITSELF

Some animals can reproduce without mating! **Starfish, flatworms, and certain lizards** can reproduce through **parthenogenesis**—a process where females create offspring **without any male involvement**. In some cases, they even produce **genetic clones** of themselves. This survival strategy kicks in when mates are scarce, allowing species to continue without traditional reproduction.

Mind-Blowing Science Fact #78

THE ANIMAL WITH A TRANSPARENT HEAD

Some fish have see-through heads!

The **barreleye fish**, found in the deep ocean, has a completely **transparent head**. Inside, you can see its **tubular eyes**, which can **rotate inside its skull** to look upward through the clear dome. This bizarre adaptation helps it spot prey in the pitch-black depths while protecting its sensitive eyes.

Mind-Blowing Science Fact #79

THE METAL THAT REMEMBERS ITS SHAPE

Some metals can "remember" their original shape!

Shape memory alloys like **Nitinol** can be bent, twisted, or deformed—and then return to their **original shape** when heated. This bizarre ability comes from the way their **crystal structure** realigns itself when exposed to heat. These metals are used in everything from **medical stents** to **self-healing eyeglass frames**.

Mind-Blowing Science Fact #80

THE INSECT THAT CAN SURVIVE DECAPITATION

Cockroaches can live without their heads!

A cockroach doesn't need its head to survive—at least, not immediately. Thanks to their simple circulatory system and the fact that they **breathe through tiny holes in their body,** a decapitated cockroach can live for up to **a week** before dying of thirst. They only die because they can't drink without a mouth!

Mind-Blowing Science Fact #81

THE SHARK THAT LIVES FOR CENTURIES

Some sharks can live over 400 years!

The **Greenland shark** holds the record for the **longest-living vertebrate** on Earth. Scientists have estimated that some individuals are over **400 years old**, meaning they were born before the **United States even existed**. These slow-moving Arctic sharks grow only about **1 cm per year** and don't reach maturity until they're around **150 years old**.

Mind-Blowing Science Fact #82

THE PLANT THAT EATS ANIMALS

Some plants are natural-born hunters! **Carnivorous plants** like **Venus flytraps, pitcher plants, and sundews** have evolved to **trap and digest insects and small animals**. They thrive in nutrient-poor soil and make up for it by catching prey with sticky traps, snap jaws, or slippery pits. It's nature flipping the food chain upside down—plants eating animals.

Mind-Blowing Science Fact #83

THE TREE THAT GROWS 40 FRUITS

One tree can grow 40 different fruits!

Artist and professor **Sam Van Aken** created the **Tree of 40 Fruit** by **grafting branches** from multiple fruit trees onto a single rootstock. The result? A tree that produces **peaches, plums, apricots, cherries, and almonds**—all from one trunk. Each branch blooms in a rainbow of colors and bears a different fruit, all thanks to clever science and horticulture.

Mind-Blowing Science Fact #84

THE SOUND THAT CAN LEVITATE OBJECTS

Sound waves can make things float!

Using a technique called **acoustic levitation**, scientists can suspend small objects like **water droplets, beads, or insects** in mid-air—**using only sound waves**. The sound's vibrations create **pressure pockets** that counteract gravity, holding the object perfectly still. It's the closest thing we have to real-life, sound-powered telekinesis.

Mind-Blowing Science Fact #85

THE RIVER THAT DISAPPEARS UNDERGROUND

Some rivers vanish into the Earth!

Disappearing rivers, also known as **losing streams**, are waterways that **suddenly vanish underground**, flowing through **limestone caves and porous rock layers**. One famous example is the **Santa Fe River** in Florida, which disappears into a sinkhole and travels underground for over **3 miles** before re-emerging. Nature's version of a magic trick.

Mind-Blowing Science Fact #86

THE LIGHTNING THAT STARTS WILDFIRES

Lightning can strike without rain!

Dry lightning occurs when thunderstorms produce lightning but **no rain reaches the ground**—often because it evaporates before falling. These invisible strikes can ignite **wildfires**, especially in dry, forested areas. In some regions, dry lightning is responsible for **up to 60% of all wildfires**.

Mind-Blowing Science Fact #87

THE ANIMAL THAT SLEEPS HALF A BRAIN

Some animals sleep with half their brain awake!

Dolphins, whales, and certain birds use a strategy called **unihemispheric slow-wave sleep**, where **one half of their brain sleeps** while the other half stays alert. This allows them to **keep swimming, breathe consciously, and watch for predators**—all while resting. They literally **sleep with one eye open.**

Mind-Blowing
Science Fact #88

THE ANIMAL THAT CAN LIVE WITHOUT OXYGEN

Some animals survive without oxygen at all! The tiny **Henneguya salminicola**, a microscopic parasite found in salmon, is the **first known animal** that **doesn't require oxygen to live**. It lacks the genes for **mitochondria**, the part of cells that usually use oxygen to generate energy. Instead, it survives entirely through **anaerobic metabolism**—proof that life can exist in ways we never imagined.

Mind-Blowing Science Fact #89

THE ANIMAL THAT SHOOTS BOILING WATER

Some creatures weaponize hot chemicals! The **bombardier beetle** defends itself by blasting predators with a spray of **boiling, noxious chemicals**. It mixes **hydrogen peroxide and quinones** inside its abdomen, then releases them in an explosive chemical reaction that reaches temperatures of **212°F (100°C)**. Nature's very own tiny chemical cannon.

Mind-Blowing Science Fact #90

THE ANIMAL THAT CAN SEE POLARIZED LIGHT

Some animals see what humans can't!

Mantis shrimp have some of the most complex eyes in the animal kingdom. They can detect **polarized light**, an invisible pattern of light waves that humans can't naturally see. This super-vision helps them **hunt prey, spot predators, and communicate** in ways we can barely imagine. Their eyes have **16 types of color receptors**—compared to our three!

Mind-Blowing Science Fact #91

THE SAND THAT SINGS

Some sand dunes can sing or boom!

In deserts around the world, certain sand dunes produce a deep, booming sound when the sand **slides or avalanches**—a phenomenon known as **singing sand** or **booming dunes**. The sound can last for **minutes** and is caused by the synchronized movement of sand grains, creating a natural, eerie hum that can be heard from miles away.

Mind-Blowing Science Fact #92

THE BACTERIA THAT EAT PLASTIC

Some bacteria can digest plastic waste!

In 2016, scientists discovered a species called **Ideonella sakaiensis** that can **break down PET plastic**, one of the most common plastics used in bottles and packaging. These bacteria produce enzymes that **decompose plastic into harmless byproducts**, offering a possible biological solution to the global plastic pollution crisis.

Mind-Blowing Science Fact #93

THE FIRE THAT BURNS WITHOUT OXYGEN

Some fires don't need air to burn!

Thermite reactions produce a chemical fire so intense it can **melt steel**—and it doesn't require oxygen to burn. The reaction uses **aluminum powder and iron oxide**, creating temperatures over **4,000°F (2,200°C)**. Because it generates its own oxygen from the chemical process, this fire can even **burn underwater**.

Mind-Blowing Science Fact #94

THE PLANT THAT MOVES ON ITS OWN

Some plants can move without wind!

The **Mimosa pudica**, also known as the **sensitive plant**, has leaves that **fold inward and droop** when touched, shaken, or exposed to heat. This rapid movement is caused by changes in **water pressure inside its cells**—a natural defense mechanism to scare off herbivores and protect itself from harm. It's one of the few plants that reacts in real time.

Mind-Blowing Science Fact #95

THE SNOW THAT'S PINK

Some snow is naturally pink!

Known as **watermelon snow**, this phenomenon occurs when microscopic **green algae** called **Chlamydomonas nivalis** bloom on snowfields and glaciers. The algae contain a red pigment that tints the snow **bright pink** and gives off a faint **watermelon scent**. It's beautiful—but it also accelerates melting by **absorbing more sunlight.**

Mind-Blowing Science Fact #96

THE OCEAN THAT GLOWS AT NIGHT

Some oceans light up like stars!

Bioluminescent plankton, like **dinoflagellates**, can make entire coastlines **glow bright blue at night** when disturbed by waves, boats, or swimmers. This glowing effect is caused by a chemical reaction inside the plankton, producing light as a defense mechanism. Certain beaches in places like **Puerto Rico** and **Maldives** are famous for this **living light show.**

Mind-Blowing Science Fact #97

THE PLANET WITH A METHANE OCEAN

Some moons have lakes made of liquid gas! Saturn's largest moon, **Titan**, has vast lakes and rivers—but they're not filled with water. Instead, they're made of **liquid methane and ethane** because of Titan's freezing temperatures. It's the only place in our solar system known to have **stable liquid on its surface**, other than Earth—except on Titan, it's a frigid sea of natural gas.

Mind-Blowing Science Fact #98

THE BUTTERFLY THAT CAN SEE UV LIGHT

B utterflies can see hidden colors! Many butterfly species have eyes that can detect **ultraviolet (UV) light**, revealing patterns on flowers and even on each other's wings that are **invisible to humans**. These UV patterns help butterflies find nectar, avoid predators, and choose mates. Their world is filled with colors and signals we'll never naturally see.

Mind-Blowing Science Fact #99

THE RIVER THAT BOILS WITHOUT FIRE

There's a river powered by heat from below!

In Peru's **Amazon rainforest**, the Shanay-Timpishka—also known as the **Boiling River**—reaches temperatures up to **200°F (93°C)**. But unlike most hot rivers, it's not heated by a nearby volcano. Instead, the river is warmed by **geothermal heat** from deep underground, creating one of the world's largest naturally boiling waterways.

Mind-Blowing Science Fact #100

THE BLACK HOLE THAT SPINS AT WARP SPEED

Some black holes spin faster than light!

The supermassive black hole at the center of the **M87 galaxy** is spinning at **near light-speed**—so fast, it's warped the fabric of space-time around it. This astonishing spin helps the black hole **generate massive jets of energy** that shoot out at near light-speed, pushing matter out across vast distances, like cosmic fire hoses in space.

CONCLUSION

Congratulations! You've just experienced *100 Mind-Blowing Science Facts* and ventured into a world of mind-bending discoveries that make our universe so incredibly fascinating. From jaw-dropping phenomena to unexpected revelations, this collection has shown that science is much more than just facts and figures—it's an endless adventure of discovery waiting to unfold.

But here's the thing about science—it's a constantly evolving journey. For every story you've read, there are countless others out there, each adding a new layer to our understanding of the world around us. Maybe this book has sparked a deeper curiosity for the wonders of nature, or perhaps it's opened your eyes to the strange and unexpected marvels of the universe. Or maybe it's simply reminded you of why science is so captivating—its ability to surprise and inspire awe with every discovery.

The truth is, the world of science is full of astonishing stories, and you don't have to be in a lab to uncover them. All it takes is a curious mind, a thirst for knowledge, and a willingness to ask, "What's next?"

So as you close this book, don't think of it as the end. Think of it as a launchpad for further exploration—because the greatest scientific wonders are still out there waiting to be discovered.

Until next time, stay curious, stay adventurous, and remember: the most mind-blowing discoveries are the ones we haven't made yet.

ACKNOWLEDGEMENTS

Creating *100 Mind-Blowing Science Facts* has been an exhilarating journey of discovery, curiosity, and countless "Wait, really?" moments. While my name may be on the cover, this book wouldn't have come to life without the inspiration, support, and contributions of some incredible people.

First, a huge thank you to all the scientists, educators, and curious minds who have dedicated their lives to exploring the wonders of our world. Your passion for unraveling the mysteries of nature and your tireless pursuit of knowledge has been the spark for so many of the amazing facts in this book. This collection is a celebration of your work and the groundbreaking discoveries that continue to shape our understanding of the universe.

To my family and friends, who've patiently listened to my endless rants about the mind-boggling ways nature works—your support and enthusiasm (and willingness to indulge my excitement over quantum physics and deep-sea creatures) have been invaluable. You kept me grounded and motivated through every step of this adventure.

A big thank you to my readers—you're the ones who make all this possible. Whether you're here to be amazed, to find fun conversation starters, or to explore the wildest corners of science, this book is for you. Your insatiable curiosity and thirst for knowledge are what keep the wonders of science alive.

And finally, to science itself—thank you for being so endlessly fascinating, mysterious, and unpredictable. Every discovery you make only raises more questions, and that's what makes you so incredibly exciting. I'm grateful for the opportunity to share just a glimpse of the awe-inspiring world you continue to reveal.

Here's to science, to the stories still waiting to be uncovered, and to the limitless wonders of the universe.

ABOUT THE AUTHOR

Felix Grayson is a storyteller at heart, driven by an insatiable curiosity for the strange, surprising, and downright unpredictable moments in science. With a passion for uncovering the most mind-boggling and awe-inspiring facts, Felix has crafted *100 Mind-Blowing Science Facts* to entertain, amaze, and spark wonder in curious minds of all ages.

When he's not digging through research or chasing down the next scientific marvel, Felix enjoys exploring museums, devouring science books, and pondering life's most fascinating mysteries over a cup of coffee. A firm believer in the power of curiosity and the magic of discovery, Felix invites you to take this journey through the world of science, proving that the universe is just as full of surprises as it is full of stars.

www.ingramcontent.com/pod-product-compliance
Lightning Source LLC
Chambersburg PA
CBHW031311120626
46554CB00001BA/359